KU-634-470

PERSONAL PROTECTIVE EQUIPMENT

PERSONAL PROTECTIVE EQUIPMENT

A Gower Health and Safety Workbook

Graham Roberts-Phelps

Gower

Published by
Gower Publishing Limited
Gower House
Croft Road
Aldershot
Hampshire GU11 3HR
England

Gower
Old Post Road
Brookfield
Vermont 05036
USA

Graham Roberts-Phelps has asserted his right under the Copyright, Designs and Patents Act 1988 to be identified as the author of this work.

British Library Cataloguing in Publication Data
Roberts-Phelps, Graham
    Personal protective equipment. – (A Gower health and safety workbook)
    1.Protective clothing 2.Protective clothing – Law and legislation – Great
    Britain
    I.Title
    687.1'6

ISBN 0 566 08061 3

Typeset in Times by Wearset, Boldon, Tyne and Wear and printed in Great Britain by print in black, Midsomer Norton.

Contents

Chapter 1
Introduction

This first chapter acts as a record of your progress through the workbook and provides a place to summarize your notes and ideas on applying or implementing any of the points covered.

PERSONAL DETAILS

Name:
Position:
Location:
Date started: Date completed:

Chapter title	Signed	Date
1. Introduction		
2. PPE: an introduction to the regulations		
3. Managing work equipment and PPE		
4. The provision and use of work equipment		
5. Key point summary		
6. Learning review		
Demonstration of PPE safety in the workplace		
Steps taken to reduce risks and hazards		

Safety review dates	Assessed by	Date
1 month	_____	_____
2 months	_____	_____
3 months	_____	_____
6 months	_____	_____

HOW TO USE THIS SELF-STUDY WORKBOOK

Overview

This self-study workbook is designed to be either one, or a combination, of the following:

- ◆ a self-study workbook to be completed during working hours in the student's normal place of work, with a review by a trainer, manager or safety officer at a later date

- ◆ a training programme workbook that can be either fully or partly completed during a training event or events, with the uncompleted sections finished in the student's normal working hours away from the training room.

It contains six self-contained chapters which should each take about 20 minutes to complete, with the final section, 'Learning Review', taking slightly longer due to the testing and validation instruments.

It is essential that students discuss their notes and answers from all sections with a supervisor, trainer or coach.

NOTES FOR TRAINERS AND MANAGERS

For use in a training session

If you are using the workbook in a training event you might choose to send the manual to students in advance of their attendance, asking them to complete the Introduction (Chapter 1). Other exercises can then be utilized as required during the programme.

For use as an open-learning or self-study tool

Make sure that you have read the workbook properly yourself and know what the answers are. Anticipate any areas where students may require further support or clarification.

Comprehension testing

Each section features one or more summary exercises to aid understanding and test retention. The final chapter, 'Learning Review', contains a set of tests, case studies and exercises that test application and knowledge. Suggested answers to these are given in the Appendix.

If you are sending the workbook out to trainees, it is advisable to send an accompanying note reproducing, or drawing attention to, the points contained in the section 'Notes for Students'. Also, be sure to set a time deadline for completing the workbook, perhaps setting a review date in advance.

The tests contained in the learning review can be marked and scored as a percentage if required. You might choose to set a 'pass' or 'fail' standard for completion of the workbook, with certification for all those attaining a suitable standard. Trainees who do not reach the required grade on first completion might then be further coached and have points discussed on an individual basis.

NOTES FOR STUDENTS

This self-study workbook is designed to help you better understand and apply the topic of safe manual handling. It may be used either as part of a training programme, or for self-study at your normal place of work, or as a combination of both.

Here are some guidelines on how to gain the most from this workbook.

- ◆ Find 20 minutes during which you will not be disturbed.

- ◆ Read, complete and review one chapter at a time.

- ◆ Do not rush any chapter or exercise – time taken now will pay dividends later.

- ◆ Complete each written exercise as fully as you can.

- ◆ Make notes of questions or points that come to mind when reading through the sections.

- ◆ Discuss anything that you do not understand with your manager, safety officer or work colleagues.

The final chapter, 'Learning Review', is a scored test that may carry a pass or fail mark.

At regular intervals throughout the workbook there are exercises to complete and opportunities to make notes on any topics or points that you feel are particularly important or relevant to you. These are marked as:

Notes

_____

_____

_____

_____

LEARNING DIARY

Personal Learning Diary

Name: _____

Job Title: _____

Company: _____

Date: _____

The value of the training programme will be greatly enhanced if you complete and review the following Learning Diary before, during and after reviewing and reading the workbook.

LEARNING OBJECTIVES

At the start or before completing the workbook, please take time to consider what you would like to learn or be able to do better as a result of the training process. Please be as specific as possible, relating points directly to the requirements of your job or work situation. If possible, please involve your manager, supervisor or team leader in agreeing these objectives.

Record these objectives below

1.

2.

3.

4.

5.

6.

> **PLEASE COMPLETE**
> **BEFORE CONTINUING**

LEARNING LOG

During the training programme there will be many useful ideas and learning points that you will want to apply in the workplace.

Key ideas/learning points	How I will apply these at work

PLEASE COMPLETE
BEFORE CONTINUING

LEARNING APPLICATION

At the end of each chapter, please consider and identify the specific opportunities for applying the skills, knowledge, behaviours and attitudes and record these below.

Action planned, with dates	Review/comments

Remember, it may take time and practice to achieve new results.
Review these goals periodically and discuss with your manager.

PLEASE COMPLETE
BEFORE CONTINUING

HOW TO GET THE BEST RESULTS FROM THIS WORKBOOK

The format of this workbook is interactive; it requires you to complete various written exercises. This aids both learning retention and comprehension and, most importantly, acts as a permanent record of completion and learning. It is therefore essential that you **complete all exercises, assignments and questions**.

In order to gain the maximum value and benefit from the time that you invest in completing this workbook, use the following guidelines.

Pace yourself

You might choose to work through the whole workbook in one session or, alternatively, you might find it easier to take one chapter at a time. This is the recommended approach. If you are using this workbook as part of a live training programme, then make time to follow through any unfinished exercises or topics afterwards.

Share your own opinions and experience

We all have a different view of the world, and we all have different backgrounds and experiences. As you complete the workbook it is essential that you relate learning points directly to your own situation, beliefs and work environment.

Please complete the exercises using relevant examples that are personal and specific to you.

Keep an open mind

Some of the material you will be covering may be simple common sense, and some of it will be familiar to you. Other ideas may not be so familiar, so it pays to keep an open mind, as most learning involves some form of change. This may take the form of changing your ideas, changing an attitude, changing your perception of what is true, or changing your behaviours and the way you do things.

When we experience change, in almost anything, our first automatic reaction is resistance, but this is not usually the most useful response. Remember, safety is something we have been aware of for a long time, and consider (or fail to consider, as the case may be!) every day. As a result, we follow procedures without thinking – on auto-pilot as it were. This often means that we have a number of bad habits of which we are unaware.

Example of change:

Sign your name here as you would normally do:

Now hold the pen or pencil in the **opposite** hand to that which you normally use and sign your name again:

Apart from noting how difficult this might have been, consider also how 'strange' and uncomfortable this seemed. You could easily learn to sign your name with either hand, usually far more quickly than you might think. However the resistance to change may take longer to overcome.

Make Notes

Making notes not only gives you information to refer to later, perhaps while reviewing the workbook, but it also aids memory. Many people find that making notes actually helps them to remember things more accurately and for longer. So, as you come across points that are particularly useful or of particular interest, please take a couple of moments to write these down, underline them or make comments in the margin or spaces provided.

Review with others

In particular, ask questions and discuss your answers and thoughts with your colleagues and fellow managers, especially points which you are not sure of, points which you are not quite clear on, and perhaps points about which you would like to understand more.

Before you start any of the main chapters, please complete the following learning assignments.

LEARNING OBJECTIVES

It is often said that if you do not know where you are
going, any road will get you there. To put it another way,
it is difficult to hit the target you cannot see. To gain the
most benefit from this workbook, it is best to have some
objectives.

Overall objectives

- **Improvements.** We don't have to be ill to improve our fitness.
  Improvement is always possible.

- **Skills.** Learn new skills, tips and techniques.

- **Knowledge.** Gain a better understanding of safety issues.

- **Attitudes.** Change the way we think about safety issues.

- **Changes.** Change specific attitudes on behaviours and our safety
  procedures and practice.

- **Ideas.** Share ideas.

Here are some areas in which you can apply your overall objectives.

1.   Hazards and risks

The first objective is to be able to identify safety hazards and risks. These may
exist all around us and may not be readily identifiable as such – for example,
the ordinary moving of boxes or small items, using a kettle or hand drill,
cleaning and so on.

2.   Prevention

Prevention is always better than cure, and part of this workbook will deal with
knowing how to prevent accidents and injuries in the first place. Injuries are
nearly always painful both in human and business terms. As well as accidents
that cause us or others harm, there are many more accidents that cause damage
and cost money to put right.

3.   Understanding your safety responsibilities

Health and safety is everybody's responsibility, and safety is a full-time job. As
you complete this workbook you will be looking at how it affects you personally
and the role that you can play, not only for your own safety but also for the safety
of others around you.

4. Identifying ways to make your workplace safer

A workbook like this also gives us the opportunity to put ideas together on how we can improve the health and safety environment of our workplace. We do not have to have safety problems in order to improve safety, any more than we have to be ill to become fitter.

An improvement in working conditions does not have to cost much or be very complicated.

Make a note here of any personal objectives that you may have.

Notes

_____

_____

_____

_____

SELF-ASSESSMENT WORKSHEET

Please complete the following questionnaire, as honestly and accurately as you can. Rate your response to each statement or question on the following scale:

1 = Never; 2 = Sometimes; 3 = Usually; 4 = Often; 5 = Always.

1. I always use personal protective equipment whenever it is available or needed	1 2 3 4 5
2. I encourage co-workers to take personal protective equipment seriously	1 2 3 4 5
3. I keep alert for personal protective equipment in bad condition	1 2 3 4 5
4. I use safety guards and shields on machines and tools	1 2 3 4 5
5. I pace myself and avoid becoming too rushed or tired	1 2 3 4 5
6. I am careful when using chemicals and other hazardous substances	1 2 3 4 5
7. I allow time for making checks before I start something new	1 2 3 4 5
8. I ask for help or advice when I am not sure of something	1 2 3 4 5
9. I can accept constructive criticism about safety practices and my work	1 2 3 4 5
10. I input my ideas into the safety planning and programmes	1 2 3 4 5
11. I am particularly cautious when dealing with glass and sharp objects	1 2 3 4 5
12. I avoid taking short-cuts and cutting corners that may involve increasing the risk of an accident or injury	1 2 3 4 5
13. I make sure that the area in which I am working is organized and tidy	1 2 3 4 5

Cont'd

14.	I make sure that my personal protective equipment is always correctly fitted and adjusted	1 2 3 4 5
15.	I avoid taking risks with other people's safety	1 2 3 4 5
16.	I am particularly careful to wear protective shoes when lifting or moving objects	1 2 3 4 5
17.	I am aware of the consequences, both legal and safety, of not wearing personal protective equipment	1 2 3 4 5
18.	I always wear ear protection whenever I am in a noisy environment	1 2 3 4 5
19.	I check all personal protective equipment regularly for cracks, damage and wear and tear	1 2 3 4 5
20.	I always use a seat belt when driving or as a passenger in a car	1 2 3 4 5

My score is _____/100   or _____%

Analysis

Between 80%–100%	Excellent.
Between 60%–80%	Very good, a really high level of safety awareness.
Between 40%–60%	OK, but there is room for improvement.
Less than 40%	An accident waiting to happen!

Make a note of those points which you need to concentrate on.

Notes

_____

_____

_____

PLEASE COMPLETE
BEFORE CONTINUING

Chapter 2
PPE: An
Introduction to
the Regulations

This chapter introduces the current regulations and standards of safe working practice with regard to the use of personal protective equipment (PPE).

Before starting this chapter, please take a few moments to make a note of any ideas or actions in the learning diary and log in Chapter 1.

OVERVIEW

The Personal Protective Equipment (PPE) at Work Regulations 1992, which came into force on 1 January 1993, are part of a series of Health and Safety regulations implementing EC Directives. They replace a number of old and often excessively detailed laws. The purpose of the PPE at Work Regulations is to ensure that certain basic duties governing the position and use of PPE apply to all situations where PPE is required, and they follow sound principles for the effective and economical use of PPE, which all employers should follow.

This chapter explains what the Regulations require and gives advice on how you can meet these requirements. It is not intended to be a definitive statement of the law.

What is PPE?

PPE is defined in the Regulations as 'all equipment (including clothing affording protection against the weather) which is intended to be worn or held by a person at work and which protects him against one or more risks to his health and safety' – for example, safety helmets, gloves, eye protection, high visibility clothing, safety footwear and safety harnesses.

Waterproof, weatherproof or insulated clothing is subject to the Regulations only if its use is necessary to protect employees against adverse climatic conditions that could otherwise adversely affect their health or safety.

A few types of equipment are not covered by the Regulations, mainly ordinary working clothes and uniforms that do not specifically protect against risks to health and safety, and protective equipment worn by professional sportspeople during competition.

Provision and use of PPE

The principal requirement of the PPE at Work Regulations 1992 is that personal protective equipment is to be supplied and used at work wherever there are risks to health and safety that cannot be adequately controlled in other ways.

Because the effectiveness of PPE can be easily compromised – for example, by not being worn properly – it should always be considered as the last resort and used only where other precautions cannot adequately reduce the risk of injury. However, where PPE is the only effective means of controlling the risks of injury or ill-health, employers must ensure that it is available for use at work – free of charge.

The self-employed

The self-employed also have a duty to obtain and use the appropriate PPE wherever there is a risk to their health and safety that cannot be adequately controlled by alternative measures.

Assessing suitable PPE

In order to choose the right type of PPE, the different hazards in the workplace need to be considered carefully. This will enable an assessment to be made of which types of PPE are suitable to protect against the hazard and for the job to be done. Your supplier should be able to advise you on the different types of PPE available and their suitability for different tasks. It may be necessary in a few particularly difficult cases to obtain advice from specialist sources – and, of course, from the PPE manufacturer.

Any PPE selected should be 'CE-marked'. This signifies that the PPE satisfies certain basic safety requirements and, in most cases, will have been tested and certified by an independent body. This is also a requirement of the Regulations.

Make a note of the points from this section that concern you.

Notes

_____

_____

_____

_____

_____

_____

_____

ASSESSING PPE

The following factors should be considered when assessing the suitability of PPE.

- Is it appropriate for the risks involved and the conditions at the place where exposure to the risk may occur? For example, eye protection designed for providing protection against agricultural pesticides will not offer adequate face protection for someone using an angle grinder to cut steel or stone.

- Does it prevent or adequately control the risks involved without increasing the overall level of risk?

- Can it be adjusted to fit the wearer correctly?

- Has the state of health of those who will be wearing it been taken into account?

- What are the needs of the job and the demands it places on the wearer? For example, how long does the PPE need to be worn, what is the physical effort required to do the job and what are the requirements for visibility and communication?

- If more than one item of PPE is being worn, are they compatible? For example, does the use of a particular type of respirator make it difficult to make eye protection fit properly?

Make a note of the points from this section that concern you.

Notes

_____

_____

_____

_____

_____

TRAINING

Make sure that the user is aware of why PPE is needed, when it is to be used, repaired or replaced and its limitations. Instruct, train and supervise its use. Because PPE is the last resort after other methods of protection have been considered users **must** wear it all the time they are exposed to the risk. Never allow exemptions for those jobs which take 'just a few minutes'. Regularly check the use of PPE and investigate fully any reasons for non-use. Use safety signs to remind people to wear PPE.

Maintenance

Equipment must be well looked after and properly accommodated when not in use – for example, stored in a dry, clean cupboard or, in the case of smaller items such as eye protection, in a box or case. It should be kept clean and in good repair – the manufacturer's maintenance schedule (including recommended replacement periods and shelf lives) should normally be followed. Simple maintenance can be carried out by the trained wearer, but more intricate repairs should only be undertaken by specialist personnel.

To avoid unnecessary loss of time, suitable replacement PPE should always be readily available.

Other Regulations

The PPE at Work Regulations do not apply where PPE is provided under seven sets of other Regulations which also require the use of some types of PPE to protect against certain risks. The seven sets of Regulations are:

- ◆ Control of Lead at Work Regulations 1980

- ◆ Ionising Radiation Regulations 1985

- ◆ Control of Asbestos at Work Regulations 1987

- ◆ Control of Substances Hazardous to Health Regulations 1994 (COSHH)

- ◆ Construction (Head Protection) Regulations 1989

- ◆ Noise at Work Regulations 1989

- ◆ Construction (Health and Safety Welfare) Regulations 1996.

Make a note of any points from this section that concern you.

Notes

_____

_____

_____

_____

_____

_____

_____

KEY POINTS TO REMEMBER

Consider whether there are ways (other than PPE) in which the risk can be adequately controlled – for example, by engineering controls.

If not, check that:

- ◆ **PPE is provided**

- ◆ **it offers adequate protection for its intended use**

- ◆ **those using it are adequately trained in its safe use**

- ◆ **it is properly maintained and defects reported**

- ◆ **it is returned to its proper accommodation after use.**

Make a note of any points from this section that concern you.

Notes

_____

_____

_____

_____

_____

_____

_____

_____

SELECTING THE RIGHT PERSONAL PROTECTIVE EQUIPMENT

Make a list of **everything** that you might consider when selecting the right personal protective equipment for your job or your work environment.

You might find it useful to divide your list into the four categories below.

Physical hazards (heat, cold, cuts, and so on)	Chemicals
Noise	General health

PLEASE COMPLETE
BEFORE CONTINUING

Chapter 3
Managing Work Equipment and PPE

This chapter shows you how to implement the Regulations.

Before starting this chapter, please take a few moments to make a note of any ideas or actions in the learning diary and log in Chapter 1.

HOW TO IMPLEMENT A PPE PROGRAMME

Equipment used at work can be the source of many varied hazards, and its misuse results in a substantial number of accidents and injuries every year. Under the Personal Protective Equipment at Work Regulations 1992 employers should regularly monitor and review their PPE programme.

Although the term 'equipment' has a broad meaning, for our purposes it is taken to cover work equipment (such as machinery) and personal protective equipment (such as gloves). In this context, it does not cover equipment provided for use in an emergency (such as fire extinguishers or fire suppression systems) and safety signs – both of which are covered by specific legislation and standards.

As with all other aspects of Health and Safety, employers are required to satisfy general duties of the Health and Safety at Work Act 1974 (HASAWA). Under this legislation, training and information should relate to the hazards, the correct use of any protective equipment issued and statutory obligations. Supervisors should also receive this training so that they can oversee the correct use of the equipment. Records should be kept of any training or information provided. Also contained within the Act are requirements for maintenance, replacement and cleaning of PPE where appropriate; and for work equipment.

More specific Regulations which must be complied with are as follows:

- Abrasive Wheels Regulations 1970

- Woodworking Machines Regulations 1974

- Provision and Use of Equipment at Work Regulations 1992

- Control of Substances Hazardous to Health Regulations 1994 (COSHH)

- Ionising Radiation Regulations 1985

- Control of Lead at Work Regulations 1980

- Control of Asbestos at Work Regulations 1987

- Construction (Head Protection) Regulations 1989

- Construction (Health and Safety Welfare) Regulations 1996

- Noise at Work Regulations 1989.

Although this might all sound rather confusing (probably because it is!), it simply means the following:

- Any equipment must be kept clean and checked for damage at regular intervals.

- A defect reporting system should be implemented so that equipment can be maintained or replaced where necessary.

- Some forms of PPE, such as self-contained breathing apparatus, require test and inspection.

- Records of test and inspection should be maintained.

This will not only provide information relating to effectiveness, but will also be useful in progressively reducing risk through engineered solutions rather than through the use of PPE.

Make a note of any improvements you can make to your PPE programme.

Notes

_____

_____

_____

_____

_____

_____

_____

ENCOURAGING THE USE OF PPE

The purpose of this exercise is to examine one very important issue regarding personal protective equipment – encouraging and enforcing its use.

Every day, intelligent and fully informed people ignore the most basic common sense and **do not wear** personal protective equipment when they **know** that they should. Have you ever considered why this should be so?

Take time to complete the following exercise in detail.

1. *Consider carefully why people frequently **do not wear** personal protective equipment, even when they know that they should and are supposedly aware of the consequences.*

2. *Consider practical ways of encouraging people to wear or use personal protective equipment more often, if not all the time.*

PLEASE COMPLETE
BEFORE CONTINUING

Chapter 4
The Provision and
Use of Work
Equipment

This chapter considers the responsibilities of employers and employees as described by the Provision and Use of Work Equipment Regulations 1992 (PUWER).

Before starting this chapter, please take a few moments to make a note of any ideas or actions in the learning diary and log in Chapter 1.

THE PROVISION AND USE OF WORK EQUIPMENT REGULATIONS 1992 (PUWER)

The Personal Protective Equipment at Work Regulations 1992 (PPEWR) generally cover most types of PPE in the workplace. They replace many earlier statutes relating to specific industries or circumstances (for example, the Protection of Eyes Regulations 1974) and cover equipment such as safety footwear, waterproof clothing, safety helmets, gloves, high visibility clothing, eye protection and safety harnesses.

They do not extend to cover sports equipment, self-defence equipment (for example, personal alarms), monitoring equipment (for example, radiation dosimeters), offensive weapons, cycle helmets, crash helmets or motorbike leathers (unless used off the highway while at work). Nor do they cover most respiratory protective equipment, hearing protection and some other types of PPE where these items are covered by the specific requirements of the Control of Substances Hazardous to Health Regulations, the Ionising Radiation Regulations, the Control of Lead at Work Regulations, the Control of Asbestos at Work Regulations and the Construction (Head Protection) Regulations.

PUWER: what do they cover?

PUWER provides a single legal code ensuring the provision of safe work equipment and its safe use in all workplaces.

'Work equipment' includes any 'machinery, appliance, apparatus or tool (including any assembly of components) which, in order to achieve a common end, are arranged and controlled so that they function as a whole'. Diverse examples include fork-lift trucks, power presses, ladders, hand-saws, company cars and overhead projectors, but not substances (such as acids, cement and water), private cars or structural items (such as walls and stairs).

Since 1 January 1997 the earlier legislative requirements contained in the Factories Act 1961, the Offices, Shops and Railway Premises Act 1963 and other related specific regulations have been totally replaced by the PUWER requirements.

Furthermore, those European directives (and subsequent UK regulations) relating to 'product safety' which impact on employers' obligations under PUWER need to be considered as well. Here, the main requirements are contained in the (amended) Supply of Machinery (Safety) Regulations 1992 which require that most machinery supplied in the UK (including imports) must:

◆ satisfy a range of health and safety requirements

◆ be subjected (in some cases) to type examination by an approved body

◆ carry 'CE' marking and other information.

Some specific equipment is excluded from these requirements as is equipment previously used in the EU – namely:

◆ equipment intended for use outside the EU

◆ equipment first supplied in the EU before 1 January 1993.

In addition, the requirements do not apply where the risks are mainly of an electrical origin (an area covered by the Low Voltage Electrical Equipment (Safety) Regulations 1989) and where the risks are wholly or partly covered by other implemented directives.

Complying with PUWER

For employers who have carefully selected and maintained their work equipment there has been little to do since the implementation of the PUWER requirements. However, the following points may serve as a useful reminder particularly as time is elapsing in relation to some of the transitional provisions. These include:

◆ a review of the assessments made under the Management of Health and Safety at Work Regulations 1992

◆ identification of the category of the work equipment used currently within the operation

◆ correct selection of equipment to ensure that it is suitable for particular tasks and ensuring that it is used only for the intended purpose

◆ consideration of any specific risk(s) posed by the equipment

◆ provision of adequate training for all users and supervisors

◆ proper maintenance of work equipment and (where provided) keeping an up-to-date maintenance log

◆ provision of information (and if appropriate, specified written instructions) to all users and supervisors of work equipment.

Risk assessments made under these requirements would include consideration of the hazards associated with particular tasks or operations, the identification of relevant statutory provisions relating to the control of identified hazards and the current preventive or control measures in place at the time of the assessment. In relation to machinery, they may be of a general nature or more specific to certain types of equipment, tasks or activities.

Such assessments will be useful in identifying the equipment used, the necessary protective measures required and the standards to be met.

Such a process will assist in determining applicable standards and the required steps to ensure compliance. Essentially, work equipment can fall into three main categories: 'old'; 'new' and 'second-hand'; 'leased or hired'.

Compliance with legal requirements and standards (such as the Supply of Machinery (Safety) Regulations 1992) is just one aspect here. Account must also be taken of working conditions, additional risks posed by the premises in which it is to be used and any additional risks posed by its intended use.

Equipment age

For old equipment, employers should ensure compliance with statutory requirements and standards identified by the risk assessment made. They should also ensure compliance with the requirements of Regulations 1–10 of PUWER and consider the requirements of Regulations 11–24 to identify any additional action that needs to be taken – some older equipment may have to be upgraded to meet these requirements.

For new equipment, employers must ensure compliance with all PUWER requirements. They should check that it complies with European directives and, where such equipment is selected or used, the requirements of Regulations 11–24 are taken to have been satisfied.

For 'second-hand', 'hired' or 'leased' equipment employers should treat it as 'new' equipment and satisfy all PUWER requirements, but not the 'essential safety requirements' of related European directives (for example, those contained in the Supply of Machinery (Safety) Regulations 1992).

Regulations 11–24 relate to dangerous parts of machinery; exposure to specified hazards; contact work equipment; articles or substances which are likely to burn, scald or sear; work equipment controls; isolating and reconnecting procedures or systems; lighting; maintenance; and markings, warnings and warning devices.

Risk assessments will help identify specific risks (such as those associated with abrasive wheels). In these circumstances, the use of that equipment and its maintenance or repair must be restricted to those who are nominated and competent.

Comprehensible information must relate to the way in which the equipment is to be used, specified safety precautions and any emergency situations or procedures.

Such training should cover any particular work methods, any risks arising out of use, the precautions to be taken and duties under the requirements.

Maintenance procedures and frequencies may be specified by manufacturers instructions, statutory requirements or experience. Ideally equipment should be readily identifiable and records of maintenance undertaken kept.

Make a note of any points from this section which you need to consider.

Notes

COMPLYING WITH PUWER

The following points may be useful in developing and implementing a strategy to ensure compliance with the statutory requirements.

Review of any risk assessments made under the requirements of the Management of Health and Safety at Work Regulations 1992

These will help determine the circumstances and the type of PPE required. It will also be useful to review assessments made under the COSHH Regulations, Ionising Radiation Regulations, the Control of Lead at Work Regulations and the Control of Asbestos at Work Regulations.

For construction operations, specific site assessments or method, statements will help determine site rules for the wearing of safety helmets and the provision of additional PPE such as protective clothing, footwear and safety harnesses.

Assessment of the suitability of the PPE before selection

Assessments of suitability should take account of the risks to be controlled, the capability of the PPE required, any additional risks created by using the equipment and a comparison of the required capabilities with those of the PPE available. Assessments in high-risk situations or for complicated PPE should be in writing.

PPE will be suitable if:

- ◆ it is appropriate for both the risk involved and the conditions of use

- ◆ it accounts for the ergonomic considerations of the circumstances (such as health of the persons who will wear it, usage and comfort)

- ◆ it is capable of proper wearer fit

- ◆ it complies with European directives, particularly the (amended) Personal Protective Equipment (EC Directive) Regulations 1992.

Provision of suitable PPE to every employee exposed and taking all reasonable steps to ensure that it is properly used

Enforcement might require the provision of site rules or instructions and appropriate disciplinary action where compliance is not observed. Conflict will be avoided if employees are involved in the selection process.

Provision of appropriate accommodation for PPE when it is not being used

Accommodation provided should be sufficient to protect against contamination, loss or damage. This might require the provision of adequate storage rooms (fitted with drying arrangements in some circumstances), containers or pegs.

Recording of equipment issue and replacement

Employees should sign for the receipt of items of personal protective equipment and these records should be kept.

Provision of adequate and comprehensible information, instruction and training to enable PPE to be used properly

All employees should receive regular and easy-to-understand instruction and training, taking into account experience level and background. Notices and information should be prominently displayed.

Make a note of any points from this section which you need to consider.

Notes

_____

_____

_____

_____

_____

_____

_____

COMPLYING WITH PUWER

1. What are your plans to review any risk assessments made under the requirements of Management of Health and Safety at Work Regulations 1992?

2. How is PPE assessed for suitability before selection or use?

3. What steps are you taking to ensure that you and your colleagues or staff use the correct PPE and use it properly?

4. How is your PPE stored when it is not being used?

5. How do you currently keep a record of equipment issued?

6. How do you currently make available adequate and comprehensible information, instruction and training in the correct usage of PPE?

> **PLEASE COMPLETE**
> **BEFORE CONTINUING**

Chapter 5
Key Point
Summary

This chapter offers a comprehensive checklist of PPE in useful categories.

Before starting this chapter, please take a few moments to make a note of any ideas or actions in the learning diary and log in Chapter 1.

PPE: KEY POINTS

- Even where engineering controls and safe systems at work have been applied, some hazards might remain. These include potential injuries to the lungs – for instance, inhaling contaminated air, injuries to the head or feet from falling materials, injuries to the eyes from flying particles or splashes of corrosive liquids, injuries to the skin from contact with corrosive materials; and injuries to the body from extremes of heat or cold.

- Personal protective equipment is needed in these cases to reduce the risk. The Personal Protective Equipment at Work Regulations 1992 give the main requirements but other special regulations cover lead, asbestos and certain other hazardous substances. Together with noise and fall arrest equipment, they may also need to be considered.

- The purpose of personal protective equipment is only as a last resort. Wherever possible, engineering controls and safe systems of work should be used instead. If PPE is still needed, the employer must provide it free of charge.

- There are many different types of PPE available, and you must know all the types that are relevant for your work, thereby ensuring that you are using all the safety equipment that is appropriate.

- You must also understand how and why you should wear PPE, particularly in terms of protecting your own safety and that of others.

Make a note of any key points which you particularly need to consider.

Notes

_____

_____

_____

_____

_____

_____

_____

OVERVIEW

Personal protective equipment, or PPE as it is more commonly known, is any device designed to protect you personally.

Before using any protective equipment you should consider the following points.

Selection and use

You must consider the nature of the hazard to which you are exposed, and the risks of an accident or an injury occurring. The considerations may also include how long you will be exposed to these hazards and their intensity. PPE must therefore be appropriate to these.

You must choose good quality products made to a recognized standard, and one that is recognized as being safe for the purpose intended. It is important to choose equipment which fits, which suits the wearer and is comfortable to use. Any equipment that restricts vision or movement can itself increase risk and be a hazard.

Maintenance

Maintenance is also a very important aspect of PPE. Equipment must be properly looked after and stored when not in use. It must also be cleaned and kept in good repair.

Employees have a responsibility to make proper use of PPE and report its loss or destruction or any fault in it.

Remember to check that PPE is providing the protection that it is designed for, and be aware of the areas or work environments where different protective equipment is required. Safety signs should be displayed in all appropriate areas.

HEAD PROTECTION

Hazards

- Impact from falling objects
- Risk of head bumping
- Hair entanglement
- Chemical drips
- Climate or temperature extremes

Options

- Helmets
- Bump caps
- Hairnets
- Sou'westers
- Cape hoods

Notes

Some safety helmets incorporate, or can be fitted with, specifically designed breathing or hearing protection. Neck protection is also important, especially for welding and similar processes.

EAR PROTECTION

Hazards

- Impact noise

- High intensities

- High and low frequency

Options

- Ear plugs or muffs

Notes

Take advice to make sure the ear protectors fit well and are suitable for the purpose intended. Use special muffs when wearing with safety helmets.

LUNG PROTECTION

Hazards

◆ Dusts, gases, vapours

Options

◆ Disposable respirators, half-masks or full-face mask respirators fitted with a filtering cartridge or canister.

◆ Powered respirators blowing filtered air into a mask, visor, helmet or hood, fresh air equipment, breathing apparatus.

Notes

Make sure that you use the right type of equipment, that it is correctly fitted, meets all the appropriate standards and is regularly tested and maintained.

EYE PROTECTION

Hazards

- Chemical or metal splash

- Dust

- Projectiles

- Gas and vapour

- Radiation

Options

- Spectacles

- Goggles

- Face screens

- Helmets

Notes

Make sure that the eye protection wearer has the right combination of impact/dust/splash protection for the task.

HAND AND ARM PROTECTION

Hazards

- Abrasions, cuts and punctures, impact injuries

- Temperature extremes

- Skin irritation, disease or contamination

- Vibration

- Chemicals

- Electric shock

Options

- Gloves

- Gauntlets

- Mitts

- Wrist cuffs

- Armlets

Notes

Do not wear gloves or mitts when operating machines, such as bench drills, where they might become caught. Make sure that gloves give protection against all likely chemicals – some are very strong. Use skin cream after working with water or solvents. Barrier creams provide some limited protection.

ELBOW AND KNEE PROTECTION

Hazards

- Heat, cold, weather

- Chemical or metal splash

- Impact or abrasions

- Excessive wear or entanglement of clothing

Options

- Conventional or disposable overalls

- Boiler suits, warehouse coats, donkey jackets, aprons

- Chemical-resistant suits

- Thermal clothing

- Pads and protective aids

Notes

The material is important. It should be chain mail, non-flammable, antistatic, chemically impermeable and high visibility. Also, check that the padding and degree of impact protection is sufficient.

BODY PROTECTION

Hazards

- ◆ Heat, cold, weather

- ◆ Chemical or metal splash

- ◆ Spray from pressure leaks or spray guns

- ◆ Impact or abrasions

- ◆ Contaminated dust

- ◆ Excessive wear or entanglement of clothing

Options

- ◆ Conventional or disposable overalls

- ◆ Boiler suits, coats, jackets, aprons

- ◆ Chemical-resistant suits

- ◆ Thermal clothing

Notes

The material is important. It should be chain mail, non-flammable, antistatic, chemically impermeable and high visibility. Also consider using safety harnesses when working from heights and life jackets when working on or near water – in the case of, for example, construction workers, water sports instructors, dock workers, police officers and so on.

LEG PROTECTION

Hazards

- Heat, cold, weather

- Chemical or metal splash

- Spray from leaks or spray guns

- Impact or abrasions

- Contaminated dust

- Excessive wear or entanglement of clothing

Options

- Conventional or disposable overalls

- Boiler suits, warehouse coats, donkey jackets, aprons

- Chemical-resistant suits

- Thermal clothing

Notes

The material is important. It should be chain mail, non-flammable, antistatic, chemically impermeable and high visibility.

FOOT PROTECTION

Hazards

- Water

- Slips

- Electrostatic build-up

- Falling objects, heavy loads, metal and chemical splash, vehicles

Options

- Safety boots and shoes with steel toe caps

- Steel mid-sole boots

- Gaiters

- Leggings, spats and clogs

Notes

Check sole patterns, material and tread; these can help grip and prevent slips in different conditions. Also consider chemical-resistant, antistatic or electrically conductive soles. Consider ankle support and cushioning.

PPE SAFETY CHECKLIST

- Always consider using personal protective equipment whenever it is available or needed.

- Encourage co-workers to take personal protective equipment seriously.

- Keep personal protective equipment in good condition and repair.

- Always use safety guards and shields on machines and tools.

- Always pace yourself.

- Be especially careful when using chemicals and other hazardous substances.

- Check things before you start something new.

- Ask for help or advice when not sure of something.

- Accept constructive criticism about safety practices and your work.

- Input ideas into safety planning and programmes.

- Be especially cautious when dealing with glass and sharp objects.

- Don't take short-cuts and cut corners.

- Keep your work areas well organized and tidy.

- Make sure that personal protective equipment is always correctly fitted and adjusted.

- Don't take risks with other people's safety.

- Always wear protective shoes when lifting or moving objects.

- Know the consequences, both legal and safety, of not wearing personal protective equipment.

- Wear ear protection when in a noisy environment.

- Check all personal protective equipment regularly for cracks, damage and wear and tear.

- Wear a seat belt when driving or a passenger in a car.

THE PPE REGULATIONS: KEY POINTS

- Personal protective equipment is needed to reduce risk.

- The Personal Protective Equipment at Work Regulations 1992 give the main requirements but other special Regulations cover lead, asbestos and certain other hazardous substances. Noise and radiation may also need to be considered.

- Personal protective equipment should only be used as a last resort. Wherever possible, engineering controls and safe systems of work should be used instead.

- If PPE is still needed, the employer must provide it free of charge.

- There are many different types of PPE available, so you should know all the types that are relevant to your work, thereby ensuring that you are using all the safety equipment that is appropriate.

- You must understand how and why you should wear PPE, particularly in terms of protecting your own safety and that of others.

- You must consider the nature of the hazard to which you are exposed and the risks of an accident or an injury occurring. These considerations may also include how long you will be exposed to the hazard and its intensity. PPE must therefore be appropriate in all these respects.

- You must choose good quality products made to a recognized standard, and one that is recognized as being safe for the purpose intended. You must choose equipment which fits, which suits the wearer and is comfortable to use. Any equipment that restricts vision or movement can itself increase risk and be a hazard.

- Maintenance is also a very important aspect of PPE. Equipment must be properly looked after and stored when not in use. It must also be cleaned and kept in good repair.

- Employees have a responsibility to make proper use of PPE and report its loss or destruction or any fault in it.

- Remember to check that PPE is providing the protection that it is designed for, and be aware of the areas or work environments where different protective equipment is required. Safety signs should be displayed in all appropriate areas.

Chapter 6
Learning Review

This chapter serves to test your knowledge and understanding of the topics covered. You should review your answers with your safety representatives or manager.

Before starting this chapter, please take a few moments to make a note of any ideas or actions in the learning diary and log in Chapter 1.

TEST YOUR KNOWLEDGE (1)

1. List three other regulations that address specific and specialist areas of risk.

    1.

    2.

    3.

2. List three of the six main requirements of the Provision and Use of Work Equipment Regulations (PUWER) discussed so far.

    1.

    2.

    3.

3. What does PUWER stand for?

4. List three types of physical hazard.

    1.

    2.

    3.

5. List three types of personal protective equipment that could protect you against physical hazards.

    1.

    2.

    3.

Test Your Knowledge (2)

1. List three types of personal protective equipment that can be used to protect your head or face.

   1.

   2.

   3.

2. List three types of personal protective equipment that can be used to protect you against chemical hazards.

   1.

   2.

   3.

3. List three types of personal protective equipment that can be used to protect you against cuts and bruises.

   1.

   2.

   3.

4. List three things to check before putting any personal protective equipment on.

   1.

   2.

   3.

> **Please complete**
> **before continuing**

TEST YOUR KNOWLEDGE (3)

1. With regard to PPE, list three things that employers are legally bound to do.

    1.

    2.

    3.

2. With regard to PPE, list three things that employees are legally bound to do.

    1.

    2.

    3.

3. If PPE is provided, you are legally bound to wear or use it.

    TRUE/FALSE?

4. List three pieces of EU legislation that may affect you.

    1.

    2.

    3.

5. List three essentials that must be provided for in the workplace.

    1.

    2.

    3.

> **PLEASE COMPLETE**
> **BEFORE CONTINUING**

Practical Problem Solving

1. Complete the first column, marking each statement as either
    Y = definitely, N = sometimes, occasionally or not at all.

2. When you have finished, please complete the column **'What can I do about it?'**, listing several different ideas.

Y/N		What can I do about it?
	I don't always consider using personal protective equipment whenever it is available or needed, and don't always use it when I should.	
	I don't encourage co-workers to take personal protective equipment seriously.	
	Some personal protective equipment is in bad condition or needs repair.	
	I, or others, don't always use safety guards and shields on machines and tools.	
	I don't always pace myself and often become too rushed or tired.	
	I am not as careful as I should be when using chemicals and other hazardous substances.	
	I don't always have, or allow time, to check things before I start something new.	
	I don't ask for help or advice when I am not sure of something.	
	I don't easily accept constructive criticism about safety practices and my work.	
	I am not especially cautious when dealing with glass and sharp objects.	
	I often take short-cuts and cut corners that may involve increasing the risk of an accident or injury.	

Cont'd

Y/N		What can I do about it?
	My personal protective equipment is not always correctly fitted and adjusted.	
	I don't always wear protective shoes when lifting or moving objects.	
	I sometimes risk not wearing ear protection when I am in a noisy environment.	

Many of these points relate directly to the legal obligations that you, or your organization, are required to fulfil as a minimum standard of health and safety. Others are warning signs of potential risks and hazards and must be acted upon if injury or accident is to be prevented. It is important they are all resolved. Add any others that you think are missing.

PLEASE COMPLETE
BEFORE CONTINUING

CASE STUDY

Read the following case study, and decide how you would answer the question on the following page.

Jim Walker was one of those types of people that you couldn't help liking. Nothing was ever too much trouble and he never had a bad word to say about anything. This made his foreman Dave's task even more difficult.

Dave was concerned about Jim's lack of attention when it came to wearing personal protective equipment and general safety. He frequently and plainly disregarded instructions when it came to wearing of safety equipment. For instance, he only wore safety gloves when his hands got cold and refused to wear the safety boots when it was too hot. The company had invested much time and money over the past 12 months in new safety guards, protective equipment and notices, and yet the workplace didn't seem to be much safer for it.

Now Dave had begun to notice that several other of the younger lads had picked up Jim's bad habits and were also neglecting to wear goggles, shoes and other protective equipment when they were required.

Dave knew that if he didn't do something soon, it would only be a matter of time before there was an accident or injury. He had tried talking to Jim about the problem and Jim had listened and promised to be more careful, but nothing changed – at least not for more than a few days.

Dave was at a loss what to do. Should he get tough with Jim and risk upsetting him, and probably everybody else? Or should he get tough with everybody in the hope that Jim would toe the line? What other way could he solve the problem other than through confrontation?

CASE STUDY: QUESTION

If Dave asked you for advice, what would you say he should do?

PLEASE COMPLETE
BEFORE CONTINUING

DOS AND DON'TS: PERSONAL PROTECTIVE EQUIPMENT

What advice would you give to somebody regarding safety awareness and using protective equipment, considering all the different types of personal protective equipment you may use?

DOS	DON'TS

Appendix
Suggested
Answers to the
Knowledge Tests

Test your knowledge (1): suggested answers

1. Any three of the following:

    Control of Lead at Work Regulations 1980
    Ionising Radiation Regulations 1985
    Control of Asbestos at Work Regulations 1987
    Control of Substances Hazardous to Health Regulations 1994 (COSHH)
    Construction (Head Protection) Regulations 1989
    Noise at Work Regulations 1989
    Construction (Health and Safety Welfare) Regulations 1986

2. Any three of the following:

    Review of any risk assessments made under the requirements of the Management
      of Health and Safety at Work Regulations 1992
    Assessment of the suitability of the PPE before selection
    Provision of suitable PPE for every employee exposed and taking all reasonable
      steps to ensure that it is properly used.
    Provision of appropriate accommodation for PPE when it is not being used
    Recording of equipment issue and replacement
    Provision of adequate and comprehensible information, instruction and training to
      enable PPE to be used properly

3. Provision and Use of Work Equipment Regulations

4. Any three of the following:

    Broken glass
    Fumes
    Dust
    Falling or flying objects
    Cold and heat extremes
    Noise
    Lead
    Asbestos

5. Any three of the following:

    Ear muffs
    Gloves
    Body suit
    Goggles
    Visor
    Hard hat
    Protective boots
    Safety harness
    Seat belt
    Thermal clothing
    Hairnet
    Sou'wester
    Hood
    Respirator
    Face mask
    Knee and elbow pads

Test your knowledge (2): suggested answers

1. Any three of the following:

   Helmet
   Hairnet
   Full-face visor
   Goggles
   Hard hat
   Sou'wester
   Hood

2. Any three of the following:

   Chemical-resistant gloves
   Goggles
   Breathing aids
   Chemical-resistant suit
   Chemical-resistant boots

3. Any three of the following:

   Safety footwear
   Knee pads
   Elbow protectors
   Gloves

4. Any three of the following:

   Cracks
   Fitting straps
   Rips and tears
   Damage
   Special instructions

Test your knowledge (3): suggested answers

1. Any three of the following:

   Provide PPE
   Train people in its use
   Lay down procedures regarding its use
   Follow the PPE Regulations
   Look after and store PPE correctly

2. Any three of the following:

   Wear it
   Look after it
   Cooperate with safety representatives
   Follow instructions regarding use

3. True

4. Any three of the following:

   COSHH Regulations 1994
   Manual Handling Operations Regulations 1992
   Provision and Use of Work Equipment Regulations 1992
   Personal Protective Equipment at Work Regulations 1992
   Electricity at Work Regulations 1989
   Workplace (Health, Safety and Welfare) Regulations 1992

Noise at Work Regulations 1989
Management of Health and Safety at Work Regulations 1992

5. Any three of the following:

Wash and toilet facilities
Fire safety equipment
Safety procedures
Personal protective equipment

Case study

The purpose of the case study exercise is for students to apply safety knowledge and awareness.

Whilst there are no 'right' answers, students should highlight legal regulations and standards that have been broken and practical ways of enforcing these.

Lessons learnt from each situation should also be identified, both in terms of what caused the incidents and what could prevent them happening in the future.

Dos and Don'ts: suggested answers

Dos	Don'ts
◆ always wear PPE	◆ cut corners or take risks
◆ make sure it fits correctly	◆ wilfully damage or misuse PPE
◆ read all notices and safety signs	◆ be casual or lazy or not wear PPE
◆ ask if you are not sure about anything.	◆ be complacent
◆ replace or repair worn or damaged PPE	◆ think that an accident can't happen to you
◆ encourage or tell others to wear their PPE	◆ work with anyone who doesn't wear PPE

HEALTH AND SAFETY WORKBOOKS

The 10 workbooks in the series are:

Fire Safety	0 566 08059 1
Safety for Managers	0 566 08060 5
Personal Protective Equipment	0 566 08061 3
Safe Manual Handling	0 566 08062 1
Environmental Awareness	0 566 08063 X
Display Screen Equipment	0 566 08064 8
Hazardous Substances	0 566 08065 6
Risk Assessment	0 566 08066 4
Safety at Work	0 566 08067 2
Office Safety	0 566 08068 0

Complete sets of all 10 workbooks are available as are multiple copies of each single title. In each case, 10 titles or 10 copies (or multiples of the same) may be purchased for the price of eight.

Print or photocopy masters

A complete set of print or photocopy masters for all 10 workbooks is available with a licence for reproducing the materials for use within your organization.

Customized editions

Customized or badged editions of all 10 workbooks, tailored to the needs of your organization and the house-style of your learning resources, are also available.

For further details of complete sets, multiple copies, photocopy/print masters or customized editions please contact Richard Dowling in the Gower Customer Service Department on 01252 317700.